The Library of FOOD CHAINS and FOOD WEBS™

PRODUCERS

in the Food Chain

ALICE B. MCGINTY
Photography by DWIGHT KUHN

The Rosen Publishing Group's
PowerKids Press™
New York

To my mother, Linda K. Blumenthal—Alice B. McGinty
To our dear friend Meghan—Dwight Kuhn

Published in 2002 by The Rosen Publishing Group, Inc.
29 East 21st Street, New York, NY 10010

First Edition

Book Design: Maria E. Melendez
Project Editor: Emily Raabe
Photographs by © Dwight Kuhn

McGinty, Alice B.
   Producers in the food chain / Alice B. McGinty.—1st ed.
         p. cm. — (The library of food chains and food webs)
 ISBN 0-8239-5752-7 (lib. bdg.)
 1. Food chains (Ecology)—Juvenile literature. [1. Food chains (Ecology) 2. Ecology.] I. Title. II. Series.
  QH541.14 .M367 2002
  577'.16—dc21

                                                                                    00-012332

Manufactured in the United States of America

Contents

What Are Food Chains?

You are a link in an important chain called a food chain. A food chain is a chain of energy. Each member of the chain gets energy from the food it eats. When one living thing eats another, a link is formed in the food chain.

The first links in the food chain are the **producers**. Producers are plants. For example, grass is a producer.

The next links in the food chain are **consumers**. Consumers eat plants or other animals. Consumers that eat plants are called herbivores. Consumers that eat other animals are called carnivores.

The last links in the food chain are the decomposers. Decomposers are tiny organisms that break down the bodies of dead plants and animals.

PRODUCER

Producers use energy from sunlight to produce their own food.

CONSUMER

These white-tailed deer are consumers that are herbivores.

DECOMPOSER

Mushrooms are decomposers.

Everything Is Connected

Ecology is the study of how living things are linked with each other and with Earth. Ecologists know that most living things belong to more than one food chain. Like people, most animals eat many kinds of foods. For example, a fox may eat berries, fish, birds, or rabbits. Rabbits are also food for weasels, hawks, or bears. This makes foxes and rabbits a part of many food chains. Food chains that are linked form a food web.

Ecologists also study the way energy is passed along food chains. A deer, for example, has to eat many plants to get enough energy to live. A mountain lion must eat many deer to get its energy. This means that there are fewer mountain lions than deer, and fewer deer than plants.

Food webs are formed whenever a creature belongs to more than one food chain. Which creature in this food web belongs to the most food chains?

Producers Make Food

How do plants make food? The answer begins with the Sun. The Sun is a star made of hot, glowing gases. The heat and light from these gases have a lot of energy. Some of the Sun's heat and light travel through space to Earth. Plants use the energy from sunlight to make food.

Earth is made up of many different areas, such as oceans, forests, deserts, and mountains. Each of these areas is called a **biome**. Certain kinds of plants grow in each biome. These producers have an important job. They are the only members of the biome's food chains that can make their own food. Animals cannot make their own food. They depend on food from producers, or from the animals that eat producers, for their energy.

Each biome has its own producers. These producers are always the first link in any food chain.

Green Leaves on Plants

A plant's leaves can be many shapes and sizes. Most of them, though, are flat and wide. This makes it easy for the leaves to soak up sunlight.

Most plants make food inside their leaves. A leaf, like all living things, is made of **cells**. Cells are the smallest working parts of living things. Inside many of the leaf's cells is a chemical called **chlorophyll**. Chlorophyll is green. It gives leaves their green color. When sunlight shines on a leaf, some of the energy from the sunlight enters the chlorophyll. The chlorophyll uses this energy to make food. Even plants that don't look green have chlorophyll. The green sometimes is covered by other colors.

These oak tree leaves are flat and broad, to soak up as much sunlight as possible.

Trees move water through their leaves through tubes called veins (shown here).

Each of a plant's cells is protected by sturdy cell walls.

Water and Sunlight

Before a plant can make food from sunlight, it needs two other ingredients. A gas in the air, called **carbon dioxide**, is one of the ingredients the plant needs. The other ingredient that the plant needs is water. The roots of the plant soak up water from the ground. The leaf uses the energy from sunlight to turn carbon dioxide and water into sugar. This process is called **photosynthesis**. Photosynthesis comes from two Greek words which mean "to put together with light." The plant uses the sugar as food. After the plant makes sugar, there is a gas left. This gas is oxygen. The plant releases the oxygen into the air through its leaves. People and animals need that oxygen to breathe.

The leaf takes in carbon dioxide and lets out oxygen through tiny holes called stomata. The stomata are on the underside of the leaf.

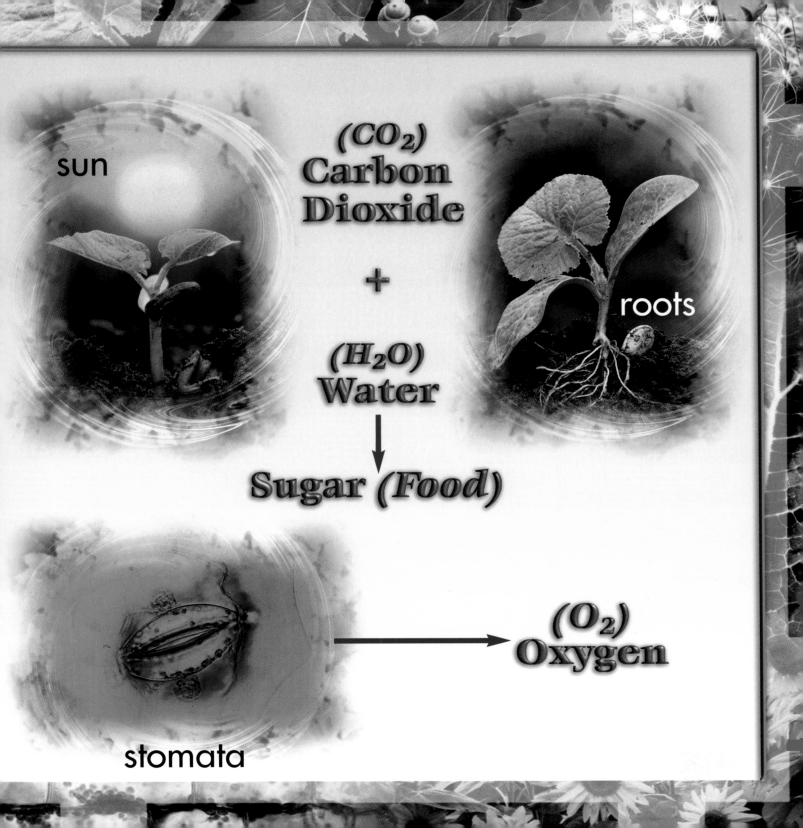

Producers and Animals

Sometimes people do things that change nature's cycles. Fuels, such as wood, gas, and coal, contain carbon. When people burn these fuels to heat homes and run cars and trucks, carbon is released into the air as carbon dioxide. Carbon dioxide acts like a blanket for Earth's atmosphere. Too much carbon dioxide will make Earth too warm.

When a plant's roots take water from the soil, they also take in **minerals**. The plant needs these minerals to grow. When an animal eats the plant, the same minerals help the animal grow. When the animal dies, decomposers return the minerals to the soil. This cycling assures that the minerals never will be used up. There are other **cycles** in the environment. Plants make oxygen and animals use that oxygen to breathe. This is called the oxygen cycle. Another cycle is the carbon dioxide cycle. When animals breathe out, they release carbon dioxide into the air. Plants use that carbon dioxide to make food.

Plants produce the oxygen that animals need to survive.

Animals use oxygen and produce carbon dioxide for plants. They also get their nutrients by eating plants.

When animals die, they rot into the soil. Plants use the nutrients from this soil to grow.

Producers in the Sea

Ocean tide pools are the areas near the shore that form pools when the tide goes out. They have their own food chains. The organisms in tide pools are special because they have to be able to live both in and out of the water. Many kinds of seaweed live in tide pools.

Almost three quarters of Earth is covered with oceans. A huge variety of creatures lives in the seas. These creatures make up countless food chains. The producers in the sea produce much of the oxygen in Earth's atmosphere. Carbon dioxide and all the nutrients the producers need to live are dissolved in the seawater. Some **algae** and tiny sea plants float free in the water. These producers are called **phytoplankton**. Phytoplankton are the first link in many ocean food chains. They are eaten by tiny animals, which may be eaten by small fish, which are eaten by bigger fish. This is another food chain!

Seaweed

The sea's producers, such as this seaweed, live near the surface of the water where sunlight shines.

Algae

Algae are one of Earth's most important producers. Many algae are tiny one-celled organisms. Other algae, such as seaweed, are bigger. Algae can be found on land, in the ocean, and in lakes and ponds.

Prickly Producers

Plants are made largely from water. Plants that live where there is little water have **adapted** to survive. The cactus plant has adapted to survive in the hot, dry desert. Cacti store water in thick stems and branches. The tall saguaro cactus can hold 1 ton (907 kg) of water in its thick stem and branches. Most cacti also do not have leaves. This is because water **evaporates** from a leaf's surface. Cacti have thick, waxy skin. This also stops the water in the cactus from evaporating. The roots of most cacti grow just under the surface of the soil. This helps them to be ready to catch dew and rain. Cacti are the first link in many desert food chains and are a good source of water.

Many cacti have thorns or spikes to protect them from being eaten.

Hot Desert

Cactus Blooming

Most cactus flowers bloom only for a day or two. Some bloom only at night. This is to prevent water from evaporating from the wide petals during hot, dry days.

Cactus Spines

Producers Who Hunt

Carnivorous plants use many tricks to trap thirsty insects. Sundews, for example, use sticky red droplets of water to trap thirsty insects. When the insects try to drink the red droplets, they become stuck.

Most land plants get the minerals they need from the soil. There are places, though, where Earth's soil does not have enough minerals for the plants to survive. Some plants that live in these areas have adapted by getting minerals in a different way. They eat insects! Plants that eat insects are called carnivorous, or meat-eating, plants. Carnivorous plants are both producers and consumers in the food chain. Carnivorous plants use water or nectar to tempt insects to land on their leaves. Then they trap the insects. The plants produce **enzymes** that dissolve the soft parts of an insect's body. The plants absorb the minerals from the insect's body.

Large, tropical pitcher plants can trap small animals, such as lizards or monkeys who put their heads inside for a drink and then get stuck.

Insects slide down the slippery leaves of the pitcher plant and are trapped in a pool of enzyme water. The insects are digested in the plant's enzyme water.

Venus's-Flytraps snap shut when an insect lands on them.

Our Lungs Need Producers

People depend on producers. We eat producers as food, and we breathe the oxygen that producers provide. Producers depend on people for the carbon dioxide we make when we breathe. Producers also depend on people to take care of them. People have cut down more than half of the world's rain forests for lumber and land. This destroys the rain forests' producers and all the other animals in their food chains. Gases from factories, power plants, and car exhaust hurt producers, too. These gases dissolve in clouds and make acid rain that damages trees. We need to protect our producers more carefully. Without producers, there would be no food chains at all!

Glossary

adapted (uh-DAP-ted) To have changed to fit new conditions.

algae (AL-jee) One-celled organisms and plants that have no roots or stems and usually live in the water.

biome (BY-ohm) An area in nature that has certain types of plants and animals.

carbon dioxide (KAR-bin dy-OK-syd) A gas that plants take in from the air and use to make food.

cells (SELZ) The many tiny units that make up living things.

chlorophyll (KLOR-oh-fill) The green chemical in producers that is used in photosynthesis.

consumers (kon-SOO-merz) Members of the food chain that eat other organisms.

cycles (SY-kulz) Series of events that are repeated in the same order.

ecology (ee-KAH-luh-jee) The study of how living things are linked with each other and with Earth.

enzymes (EN-zymz) Substances made by cells that cause changes to other substances.

evaporates (ih-VA-puh-raytz) To change from a liquid to a gas.

minerals (MIN-rulz) Natural ingredients from Earth's soil, such as coal or copper, that come from the ground and are not plants, animals, or other living things.

photosynthesis (foh-toh-SIN-thuh-sis) The process in which leaves use energy from sunlight, gases from air, and water from soil to make food and release oxygen.

phytoplankton (fy-toh-PLANK-tun) Algae and one-celled ocean plants.

producers (pruh-DOO-serz) Plants or tiny organisms that use photosynthesis to make their own food.

Index

Web Sites

To learn more about producers and food chains, check out these Web Sites:

http://passporttoknowledge.com/rainforest/main.html
www.worldwildlife.org